YOUR KNOWLEDGE HAS VALUE

- We will publish your bachelor's and
  master's thesis, essays and papers

- Your own eBook and book -
  sold worldwide in all relevant shops

- Earn money with each sale

Upload your text at www.GRIN.com
and publish for free

Sven-David Müller

Nutritional therapy for Arthritis and Osteoarthritis

Dietary therapy options for the patient with Arthritis urica (Gout), Rheumatism, Arthritis and Osteoarthritis

GRIN Publishing

Bibliographic information published by the German National Library:

The German National Library lists this publication in the National Bibliography; detailed bibliographic data are available on the Internet at http://dnb.dnb.de .

Imprint:

Copyright © 2011 GRIN Verlag GmbH
Print and binding: Books on Demand GmbH, Norderstedt Germany
ISBN: 978-3-656-84408-2

This book at GRIN:

http://www.grin.com/en/e-book/284731/nutritional-therapy-for-arthritis-and-osteoarthritis

Nutritional therapy for Arthritis and Osteoarthritis.

Dietary therapy options for the patient with Arthritis urica (Gout), Rheumatism, Arthritis and Osteoarthritis

by Sven-David Müller, M.Sc.

Cover image: morguefile.com

Musculoskeletal disorders can be of inflammatory as well as degenerative nature. 10 to 15 percent of all patients presenting at medical surgeries suffer from diseases and conditions affecting the musculoskeletal system. It is estimated that about 2.5 to 3 percent of the German population suffer from inflammatory rheumatic diseases, such as rheumatoid arthritis. These figures emphasise both, the need for providing those affected and interested in the prevention, diagnosis and treatment of inflammatory and degenerative diseases of the musculoskeletal system with up-to-date information, and the socio-economic impact and burden on the healthcare system of this group of diseases. Rheumatoid diseases involve conditions associated with pain and functional limitations of the musculoskeletal system and as such, diseases of the peripheral joints have to be differentiated from diseases affecting the root skeleton and the ones affecting connective tissues. People adversely affected by rheumatoid arthritis, are known to benefit from an anti-inflammatory nutritional therapy, as it is described in this book. This diet is low in inflammation-promoting arachidonic acid and rich in omega-3 fatty acids, which are known to reduce inflammatory reactions. At least 800.000 people in Germany suffer from rheumatoid arthritis. Osteoarthritis, also known as degenerative joint disease is characterised by the progressive degradation of cartilage. As part of the nutritional therapy, sufferers of osteoarthritis benefit from a healthy, balanced diet that helps to reduce weight and to prevent weight gain. Patients suffering from osteoarthritis often develop secondary inflammatory complications (activated osteoarthritis) and, hence, benefit from a similar diet recommended in patients suffering from rheumatoid arthritis.

Gout (Arthritis urica)

In the past, gout (hyperuricemia) predominantly affected wealthy people. Gout used to be called 'Zipperlein' and was considered to be a typical rich man's disease. Gout, today and in the past, is and used to be associated with general overeating and reduced physical activity. Nowadays, about 3 percent of men who reach the age of 65 suffer from an acute gout attack.

Definition

Gout is a rheumatic disease that is characterised by recurrent attacks. It results from an elevated concentration of uric acid in the blood, which leads to the deposition of monosodium urate crystals mainly in the joint capsules and cartilage, in the pinna of the ear and in the renal tubules. The accretions of urate crystals in the gout nodules are called tophi. Gout most often affects the big toe joint, and in those cases is referred to as podagra. The condition can be classified into primary (familial) and secondary hyperuricemia. Primary hyperuricemia is due to a congenital disorder of the purine metabolism, which in 75 to 80 % of all cases affects the excretion via the kidneys and in 20 to 25 % leads to an increase in uric acid formation. In contrast, secondary hyperuricemia is not based on a dysfunctional metabolism, but is the result of urate hypoexcretion or an elevated synthesis of uric acid. Despite many similarities with rheumatic diseases (joint pain, -inflammation, -destruction, involvement of bone, cartilage, tendons and bursae), gout classifies as a metabolic disease.

Causes

Gout is the result of a dysfunctional purine metabolism. Purines such as adenine, hypoxanthine and guanine are components of nucleic acids and thus, of the DNA or RNA, and are present in all animal and human cell nuclei. Uric acid is the end product of the human purine metabolism. The size of the body uric acid pool depends directly on both intrinsic purine synthesis (350 mg daily) and extrinsic intake of purines via the diet with more than 300 mg daily. In healthy subjects, the supply/production and the excretion of uric acid is well-balanced. This balance is disturbed in patients suffering from gout, which may result in a significantly higher total body uric acid pool than in nonhyperuricemics. Healthy individuals maintain a relatively stable serum uric acid concentration between 2 and 7 mg/dl. Is the uric acid concentration greater than 6.5 mg/l, it may promote uric acid precipitation.

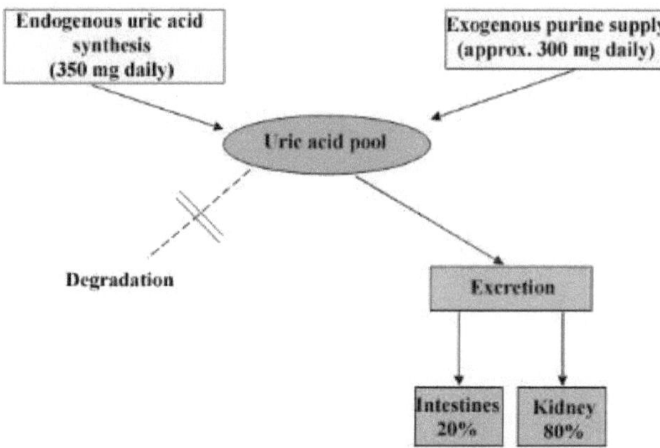

Figure: Synthesis, supply and excretion of uric acid

Gout symptoms

An acute attack of gout is characterised by the swelling and redness of the joints.
In 2/3 of all cases, the big toe joint is affected. However, this can extend to adjacent areas. Further accompanying symptoms are the general feeling of being unwell, fever, increased pulse rate, headache and vomiting. Four different types of gout can be distinguished:
1. Asymptomatic hyperuricemia: elevated uric acid levels, usually discovered by accident. The patient is usually completely asymptomatic.
2. Acute gout attack: is caused by sodium urate in the synovial space and results in significant swelling, inflammation and effusion. An acute attack commonly occurs at night or early in the morning and is extremely painful. Opulent meals together with the consumption of alcohol can trigger an attack. Characteristic is the infection of the big toe joint, but also of finger -and wrist joints and also the ankle joint may be affected.
3. Intercritical gout: the time between two attacks of gout. These periods are asymptomatic. It may take months or years until a new attack occurs.
4. Chronic (tophaceous) gout: is characterised by uric acid crystal deposits in multiple joints mediating an acute inflammatory response, as well as cartilage and bone damage with characteristic arthritic joint changes.

Diagnosis of gout

A diagnosis of hyperuricemia is based on the measurement of the uric acid level and the detection of monosodium urate crystals in the tissues and joints.

The aim of long-term treatment of gout is to permanently reduce the serum urate levels in the body. A nutrition therapy can help with achieving this goal. The overall supply of purines, which are present in animal and in plant foods as the building blocks of RNA, DNA and nucleotides, has to be reduced. Medical therapy involves the employment of pharmaceutical drugs that (a) inhibit the synthesis of uric acid (uricostatic drugs) and (b) increase the excretion of uric acid (uricosuric drugs).

Nutrition therapy of patients with hyperuricemia and gout

A consistent change of diet helps to reduce the number of prescribed pharmaceutical drugs or indeed renders them unnecessary. The frequent over- and malnutrition in western industrialised countries is considered to be one of the ultimate causes for the development of hyperuricemia and gout. The objectives of a change in diet in case of hyperuricemia are as follows:
- Restricted intake of dietary purine
- Reduction of body weight in overweight patients
- Milk and dairy products are recommended as sources of protein
- Restricted alcohol consumption

A diet low in purines should not contain more than 300 to 500 mg of uric acid per day or up to 3500 mg per week.
- Meat, meat products and fish consumption should be limited to 100 g per day. The purine-rich skin of poultry should be removed.
- Animal foods with more than 200 mg of purines per 100 g, such as offal, pork rinds, meat extract, certain fish species and all herbal products containing more than 50 mg of purines per 100 g, such as legumes, green peas, broccoli, and wheat germ should be avoided.
- Preference for low-fat dairy products and eggs are recommended as sources for animal protein.
- Offal, such as liver, kidney, thymus, heart, some types of fish and crustaceans, such as salt herring and lobster should be avoided.
- Legumes and purine-rich plant-based foods should be avoided (cabbage, Brussels sprouts, peas, beans and lentils).
- Limit alcohol consumption to one glass of wine or beer daily. 100 ml beer contains 15 mg uric acid. Non-alcoholic beer contains about the same amount of purines.
- Tea, cocoa and coffee are allowed.
- The daily fluid intake should exceed 2.5 litres. The induced diuresis results in an increase of uric acid excretion via the kidneys. Alkalising carbonated sparkling mineral waters (mediating an increase of the urinary pH) are recommended.
- It is best to cook the food, because part of the purines break down and dissolve into the cooking water.

A strict diet contains no more than 300 mg of uric acid per day or 2000 mg per week. The rules of a low-purine diet apply. In addition, meat, -sausage or fish intake should be limited to a maximum of 100 g once or twice a week.

Uric acid content in different foods per 100 g and per serving			
	Uric acid in mg per 100 g	Serving in g	Uric acid in mg per serving

Pork	150	150	225
Beef	140	150	210
Chicken legs	160	150	240
Trout	200	150	400
Beans (white)	180	50	90
Beans (green)	42	150	63
Peas	150	150	225
Brussels sprouts	60	150	90
Spinach	50	200	100
Cauliflower	45	150	68
Chinese cabbage	25	50	12
Asparagus	25	200	50
Lettuce	24	30	7

Treatment of an acute gout attack

In case of an acute gout attack, a liquid-enriched, strictly low-purine diet should be implemented. Tea, juices and alkalising mineral waters are suitable to cover the recommended fluid intake. The diet should be light and easily digestible. For the duration of the acute phase, a rice-fruit or fruit diet is recommended.

Rice-fruit diet: 250 to 300 g rice (dry weight) and 750 to 1000 g fruit are prepared for 5 to 6 meals a day without added salt, milk or fat. The selection of fruits and the type of preparation should be varied: as compote, uncooked fruit or fruit salad. To add flavour, a little bit of sugar, vanilla sugar, cinnamon or lemon additive may be used.

Fruit diet: Consists of 1250 to 1500 g different fruit five or six times daily. Avocados, dried fruit, nuts, and ripe fruit should be avoided. Sugar-rich varieties such as grapes, bananas or cherries should only be used when specifically prescribed. The fruit can be eaten fresh, as unsweetened fruit salad or, for people with sensitive stomachs, as unsweetened compote.

Rheumatoid arthritis

Already the Greek physician Hippocrates (460-377 BC) and the great German physician Paracelsus (1493-1541) were familiar with rheumatism and its treatment. Rheumatic diseases comprise a wide spectrum of different conditions. Rheumatic disorders are all musculoskeletal and connective tissue disorders. The term rheumatism is derived from Greek and means flowing, streaming. It is an outdated, inaccurate traditional term, not sufficient to describe the wide variety of rheumatic diseases, associated with flowing, raging and pulling musculoskeletal pains. Only inflammatory rheumatic diseases such as rheumatoid arthritis can be modified by nutritional therapy.

Only recently, scientific studies have established that only animal foods contain substances that promote inflammation of the joints. The inflammation is closely associated with an elevated level of arachidonic acid. It is contained only in fatty animal foods. About one in 10 adults in Germany suffers from the symptoms of rheumatic disorders. Rheumatism is a term describing painful diseases affecting the musculoskeletal system (joints, spine or muscles). In fact, rheumatism is a collective term for more than a 100 different diseases. They are all characterised by pains affecting the musculoskeletal system and by limited joint mobility. Further complications are swelling and partial or even complete loss of function of the affected body regions.

The underlying immunological mechanisms in the pathogenesis of rheumatic diseases are poorly understood and addressed in medical scientific research. In addition to hereditary factors, which play a major role in both the inflammatory as well as the degenerative rheumatic diseases, bacterial infections, stress, and chemical and physical agents may also be relevant triggers. The inflammatory reactions characteristic for inflammatory rheumatic diseases are mediated by eicosanoids and cytokines (inflammatory mediators). The most common forms of rheumatism are arthritis, osteoarthritis, soft tissue rheumatism, deterioration of the spine and chronic polyarthritis. Also diseases like gout, osteoporosis and Morbus Bechterew (Spondylitis ankylosans) belong to this group of rheumatic diseases.

The prevalence of various rheumatic diseases in the German population is as follows:

Arthritis/Osteoarthritis (Degenerative-rheumatoid diseases of the joints) affected	5 million people
Soft-tissue rheumatism (Affected are the soft parts of the musculoskeletal system)	1.6 million people affected
Chronic polyarthritis (Inflammatory autoimmune disease affecting multiple joints) affected	1 million people
Morbus Bechterew (Stiffening rheumatoid inflammation of the spine)	800 thousand affected

Treatment of rheumatic disease

Medical treatment is primarily aimed at the alleviation of clinical symptoms. The most commonly used anti-rheumatic drugs are non-steroidal anti-inflammatory drugs, like cortisone and disease-modifying anti-rheumatic drugs which predominantly target inflammation. Different types of non-steroidal anti-inflammatory drugs have different degrees of efficacy in treating rheumatic pain. Disadvantages of these pharmaceutical drugs are the relatively frequent adverse events (side effects) of partly serious nature, as well as contraindications that prohibit the use in patients.

Principles of nutrition therapy in rheumatic diseases

Nutrition therapy instead of medical treatment – a dream for many rheumatic patients struggling with side effects or suffering from severe pain, despite the use of medication. A proper diet and regular physiotherapy cannot replace the medical treatment of rheumatism, but it provides side-effect free, cheap, and delicious tasting support in the fight against chronic disease. Rheumatic diseases are not diet-related, such as obesity, gout or diabetes mellitus type 2. However, already Hippocrates described the correlation between diet and the pathogenesis of rheumatic diseases.

Rheumatoid arthritis is an autoimmune disease caused by chronic inflammation of the joints, the symptoms of which can be alleviated by nutrition therapy. Basically, it has been discovered, that patients suffering from primary chronic polyarthritis, experience alleviation of their discomforts and can potentially reduce their medication, if they eat mainly plant-based foods, low-fat animal foods, plenty of fish (at least 3-4 times a week) and use supplements like anti-oxidant vitamins, minerals, and omega-3-fatty acids. In a survey, 61 of 140 of rheumatic patients stated, that eating meat and sausage products lead to a worsening of rheumatic pains and 27 indicated that this was caused by eating animal fats and dairy products. 40 of 140 rheumatic patients observed an improvement of their symptoms after consuming vegetable food, 36 after fasting, 57 after a diet with a high ratio of raw food and 17 after eating vegetable fats.

Obesity: The enemy of the rheumatic patient

Many rheumatic patients are overweight. Every kilo too much counts and puts strain on the musculoskeletal system. Weight reduction for the obese rheumatic patient is the first and most important step to reduce pain.

How to best lose weight when suffering from rheumatic disease

After years of discussion about the appropriate way to lose weight, it has become evident that a low-fat diet will result in weight loss. Weight loss of 500 g per week is completely sufficient. During a weight-reducing diet the energy supply should ideally be 1200-1800 kilocalories. It is not the potatoes, rice, pasta, bread, bananas or grapes that are fattening the nation, but the huge meat and sausage portions, which increase the uric acid pool in the body. Unlike fats, carbohydrates are not fattening and do not contain arachidonic acid.

Carbohydrates

The body's energy supply is directly derived from carbohydrate-containing foods, such as grain products (whole grain bread, whole grain rolls, whole grain rice, and whole grain pasta), vegetables, salads, potatoes, fruit and sugar. Apart from sugar and white flour products, carbohydrate-rich foods are relatively low in calories but rich in valuable fibre.

Protein

Albumen is scientifically referred to as protein, one of the building blocks necessary for life. Protein is required for the building and repair of body tissues, including muscles. Other substances, for instance hormones and enzymes are usually protein-based. Rheumatic patients meet their protein needs by eating plant-based foods, low fat dairy products and in particular fish. Contrary to general belief, the consumption of pork has no negative effect on joint diseases or rheumatic disorders.

Fats

Fat is the most energy-rich nutrient, which is why physicians and dieticians suggest that fat makes you fat. Rheumatic patients should use only high quality vitamin E-rich vegetable oils and low fat- or low-salt margarine. Rheumatic patients, especially those suffering from obesity, benefit from an economical use of the 'good fats'.

Everyone should drink at least 2 litres a day. Most drinks have absolutely no effect on the course of rheumatic disease. Only alcoholic beverages may enhance the inflammatory process and hence, should be largely avoided. Especially recommended are black-, herbal- and fruit teas, coffee (up to 4 cups per day), nutrient-rich fruit and vegetable juice and mineral water (rich in calcium> 250 mg/l).

Inflammatory mediators

Primary chronic polyarthritis is considered to be the 'true/original arthritis'. The causes of the inflammatory processes are well-known in this type of arthritis. They are mediated by cell-derived substances (inflammatory mediators such as leukotriene B4). Nutrition therapy aims to reduce the synthesis of eicosanoid-derived inflammatory mediators. Eicosanoids are hormone-like substances, derived from 20-carbon polyunsaturated fatty acids. Therefore, scientists refer to these mediators as eicosanoids (eikos means 20 in Greek).

> Inflammatory mediators $\Rightarrow$ Inflammation $\Rightarrow$ Pain

The body's inflammatory responses play a pivotal role in the pathogenesis of rheumatic disease. Inflammatory mediators, on the other hand, are affected by our diet, such as antioxidant vitamins and minerals, and in particular the fatty acids. Particularly, eicosanoids derived from the polyunsaturated fatty acid arachidonic acid, have profound physiological effects by mediating the inflammatory response of the joints.

The Culprit: Arachidonic acid

Arachidonic acid is a polyunsaturated fatty acid which is synthesised in all mammals, including humans, and is converted from a vegetable fatty acid called linolenic acid. In addition to the endogenous biosynthesis of arachidonic acid, humans consume arachidonic acid-rich animal-derived foods. It is easier for the human body to metabolise arachidonic acid supplied via various animal food sources, rather than synthesising it from a precursor via the endogenous metabolism. Arachidonic acid is supplied exclusively and in abundance with a normal diet of animal-derived foods. All plant-based foods are free of arachidonic acid. The more arachidonic acid is available, the more pro-inflammatory eicosanoids are synthesised. Leukotriene B4 is an eicosanoid inflammatory mediator and is produced by oxidation of arachidonic acid.

> Arachidonic acid $\Rightarrow$ Inflammatory mediators $\Rightarrow$ Inflammation $\Rightarrow$ Pain

Time and again we can read and hear that fasting (a zero-calorie diet) alleviates symptoms of inflammatory rheumatic diseases (rheumatoid arthritis). According to Adam, the average daily arachidonic acid intake in Germany is 300 mg. This compares to a depletion of only 0.1 mg. In experiments, a reduction of the arachidonic acid supply/production results in lower serum uric acid levels and the synthesis of eicosanoids. The details of the arachidonic acid content in older food composition tables and nutrient calculation programs are often wrong. The data in our cookbook originate mostly from Professor Dr. Dr. Olaf Adam's research.

The arachidonic acid content in food
(mg/100 g)

Animal foods contain arachidonic acid. The lower the arachidonic acid content the better. Only fish contains in addition to arachidonic acid also abundant omega-3 fatty acids, which compensate for the adverse effect of arachidonic acid.

Milk and dairy products	
Milk, 3.5 % fat	4
Milk, 1.5 % fat	2
Milk, 0.3 % fat	0
Condensated milk, 7.5 % fat	8
Sour cream, 10 % fat	11
Whipping cream, 30 % fat	32
Buttermilk, 1 % fat	1
Natural yoghurt, 3.5 % fat	4
Natural yoghurt, 1.5 % fat	2
Whey	0
Cheese and curd cheese	
Camembert, 30 % FDM	13
Camembert, 45 % FDM	22
Camembert, 60 % FDM	34
Emmenthal cheese, 45 % FDM	28
Tilsiter, 45 % FDM	27
Regular curd cheese, 20 % fat	5
Regular curd cheese, low-fat	0
Chicken eggs	
1 chicken egg	70
Egg yolk	297
Animal fats	
Butter	83
Lard	1700
Offal	
Calf's liver	352
Pig's liver	870
Ox liver	210
Meat	
Lean pork	120
Lean beef	70
Lean veal	53
Sausage and ham	
Cooked ham	50
Smoked ham	130
Streaky bacon	250
Liver sausage	230
Pork sausage and sausages	120
Salami and cervelate sausage	100
Poultry	
Chicken breast	112
Chicken legs	190
Turkey breast	50
Turkey legs	150
Fish and sea food	
Halibut	27
Hake*	29

Thuna*	280
Herring	37
Cod	3
Mackerel	120
Redfish*	240
Sardines and anchovies	10
Haddock	2
Sole	23
Eel*	120
Trout*	30
Salmon*	300

Plant-based food products do not contain arachidonic acid. The lower the intake of arachidonic acid the better.

Plant-based oils and fats	
Plant-based oils	0
Vegetable margarine	0
Diet- and low-fat margarine	0
Vegetables, legumes, potatoes and nuts	0
Rice and egg free pasta	0
Soy products	0
Corn, flour, bread, bread rolls, egg-free pastries	0
Fruit	0
Water, tea, coffee, fruit juice and lemonade	0
Sugar, jam, honey	0

Note: * The contained omega-3 fatty acids ensure that the arachidonic acid won't have a negative impact.

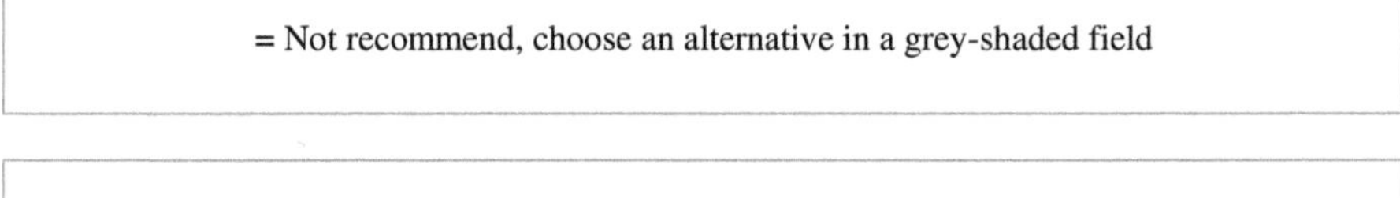

Unfortunately, the arachidonic acid content has only been analysed in the previously mentioned food products. It should be noted, that all pure vegetable foods and ready made products do not contain any traces of arachidonic acid. When it comes to animal foods, if in doubt, you can draw a comparison value from our table (for example: if the arachidonic acid content of Gouda is not analysed, you can check the arachidonic acid content of another hard cheese with the same level of fat content). The arachidonic acid content of fish should not be considered in your calculation. For the rheumatic patient a lacto-vegetarian, fish-enriched diet is recommended. The term 'lacto' stands for milk and dairy products. The word 'vegetarian' stands for plant-based food. This diet contains mostly plant-based foods, and allows consumption of low-fat milk and dairy products and fish. Meat, eggs and high-fat dairy products are largely excluded. Fish contains arachidonic acid, however, is very useful as it contains the anti-inflammatory omega-3 fatty acids for compensation (eicosapentaenoic and docosahexaenoic acid). In addition, omega-3 fatty acids reduce the formation of arachidonic acid. For inflammatory rheumatic diseases this diet forms the basis for every successful treatment, because it emphasises the consumption of plant-based food and ensures an arachidonic acid intake of generally not more than 50 mg per day. It should be noted that arachidonic acid from fish must not be taken into account, because of its benefits for rheumatism.

In acutely inflamed joints, pro-inflammatory arachidonic acid-derived eicosanoids like Thromboxane A2, Prostaglandin I2 and Leukotriene B4 are released The more of these eicosanoid mediators are available, the more severe is the inflammation. Arachidonic acid is metabolised slowly in the body and is deposited in the cell walls. Only after a few days of fasting or a vegetarian diet, arachidonic blood levels decrease.

Effects of polyunsaturated fatty acids

Humans have a limited capacity to convert linolenic into arachidonic acid, as a result, the size of the arachidonic acid pool is especially dependent on the arachidonic acid supply via food. All polyunsaturated fatty acids are inhibitors of the endogenous biosynthesis of arachidonic acid. The synthesis of arachidonic acid is tightly regulated and depends in particular on the exogenous arachidonic acid supply. A reduction of the arachidonic acid supply does not lead to an increase in the formation of arachidonic acid from different polyunsaturated fatty acids. A vegetarian diet has a number of positive effects, such as low levels of arachidonic acid and high content of polyunsaturated linolenic acid. Vegetable fats like linseed oil, rapeseed oil and soybean oil are rich in linolenic acid. Margarines made from those oils are highly recommended to the rheumatic patient.

Fasting

Time and again we can read and hear that fasting (a zero-calorie diet) positively affects inflammatory rheumatic diseases (chronic polyarthritis). Numerous studies have demonstrated, that fasting patients with chronic polyarthritis often surprisingly improve. Fasting usually means zero-calorie intake with a daily fluid intake (calorie-free) of at least 2.5 litres. Recommended is fasting with protein modified juice (1-2 litres vegetable or fruit juice + 50 g of high biological value protein from low-fat dairy products). A symptomatic improvement usually sets in within a few days. However, if eating continues as before the fasting, the condition worsens. The level of inflammation is kept at bay, if a low-fat vegetarian diet based on dairy products and fish is maintained after the fasting.

The effect of fasting

Within two days of fasting the production of eicosanoids reduces to approximately one third of the original value. The reason for this could be a decline of the eicosanoid level in the body. The drop is causally connected with the non-existent arachidonic acid supply via the food. In one fasting study, a significant improvement of symptoms was observed in both, the lab test results as well as the discomforts associated with the medical condition.

Various long-chain omega-3 fatty acids can be found in fish oil. The most important are docosapentaenoic acid, docosahexaenoic acid and eicospentaenoic acid. The eicosapentaenoic acid has great structural similarities to arachidonic acid and thereby inhibits arachidonic acid production and the synthesis of inflammatory mediators in the body. Contrary to arachidonic acid, omega-3 fatty acids have a positive effect on the inflammatory situation (especially on eicosapentaenoic acid), as they inhibit the conversion of arachidonic acid to eicosanoids and thereby eventually the formation of inflammatory mediators. However, this process takes time, a few days to a few weeks. A diet that is low in arachidonic acid and rich in omega-3 fatty acids inhibits inflammatory responses (such a diet has been reported to be effective not only in the joints - also in chronic inflammatory bowel disease, MS and other inflammatory diseases and possibly even in cases of migraines). Fish oils have an anti-inflammatory effect by inhibiting the formation of pro-inflammatory eicosanoids as well as the antioxidant activity of vitamin E. The lower the supply of arachidonic acid via the diet, the more profound is the anti-inflammatory effect of eicosapentaenoic acids found in fish oil.

During studies carried out in the 1930's on the Inuit, it was found that inflammatory rheumatic diseases in this group of people were extremely rare. This finding can be attributed directly to their diet based mainly on fish.

Particularly rich in anti-inflammatory fish oils (omega-3 fatty acids) are herring, salmon, mackerel, tuna, sardine, redfish, eel, carp, trout, halibut, and walleye. But all the other fish and other marine animals also contain valuable omega-3 fatty acids. The consumption of vegetable omega-3 fatty acids, as for example from linseed oil, rapeseed oil or walnut oil is also recommended.

Note: The required amount of fish oil for the treatment of rheumatic patients requires either at least two servings of fish per day or fish oil capsules, which are prescribed by the physician.

In a study, patients with inflammatory rheumatic diseases were given 2.7 g eicosapentaenoic acid and 1.8 g docosahexaenoic acid daily. A number of clinical symptoms improved significantly under those conditions. In particular, improved mobility of the affected rheumatic joints and a decrease in morning stiffness in the small finger joints were recorded. At the same time, a regression of parameters indicative of inflammation, including a decrease of certain inflammatory tissue reactions promoted by leukotrienes, were determined. In other studies, which included the administration of 10 grams of fish oil daily, it was found, that the need for medication (especially non-steroidal antiphlogistic = cortisone-free anti-inflammatory drugs) significantly decreased, including the degree of inflammation. The most common non-steroidal antiphlogistic agents/drugs include acetylsalicylic acid, indomethacin, diclofenac and ibuprofen. Also, cortisone is commonly used as an anti-inflammatory drug in inflammatory rheumatic diseases. On top of these, the rheumatologist often prescribes other medication, which he summarises as disease-modifying anti-rheumatic drugs. These include gold preparations, methotrexate, sulfasalazine, azathioprine, penicillamine and chloroquine.

So-called 'Rheumatism diets'

Diets that have been recommended for many years as dietary therapy in inflammatory rheumatic diseases and that are not low in arachidonic acid and rich in highly unsaturated fatty acids, do not withstand a critical clinical examination. It has frequently been suggested, to avoid foods containing additives such as preservatives and fruit, red meat, dairy products and spices. An exact study showed clearly that these dietary recommendations are useless for the rheumatic patient.

Vitamins and minerals in inflammatory rheumatic diseases

The processes underlying the increased production of inflammatory mediators are also affected by antioxidant vitamins and antioxidant enzyme systems. For the synthesis of these enzymes, trace elements like iron, zinc and selenium are essential. An optimal supply of the body with vitamins A, E and C as well as the trace elements selenium and zinc reduces the production of inflammatory mediators. As a result of chronic inflammation in inflammatory rheumatic diseases, the need for antioxidants in rheumatic patients is significantly higher than in healthy subjects. 50 to 60 percent of patients suffering from rheumatoid arthritis are inadequately supplied with vitamin E. Studies have shown, that an increased demand of antioxidants cannot be covered via the diet. The use of vitamin and mineral supplements as drugs is apparently required. Vitamin C, selenium, and zinc are thought to have an anti-inflammatory effect.

Vitamin E inhibits inflammation

Vitamin E is a lipid-soluble compound. Vitamin E contained in food is significantly more effective than synthetic vitamin E. Rheumatic patients should therefore choose vitamin E supplements of natural origin. The vitamin E of these supplements is extracted from food sources. It inhibits the arachidonic acid synthesis and the formation of inflammatory mediators. It also counteracts other immunological effects of inflammation. In rheumatic patients, the vitamin E plasma levels are significantly reduced (50-60 %). Patients with inflammatory rheumatic diseases are undersupplied with vitamin E. In studies, usually at least 30 % of rheumatic patients suffering from chronic polyarthritis are also affected by a vitamin E deficiency. This reinforces substitution therapy with vitamin E. Inflammatory processes affecting the joints may lead to a local deficiency of vitamin E. The vitamin E concentration in the synovial fluid often drastically decreases by about 4/5, compared to the vitamin E concentration in the blood. And also this concentration is often significantly decreased compared to healthy subjects.

Food products rich in vitamin E
Corn oil
Rapeseed oil
Sunflower oil
Wheat germ oil (10 g cover 133 % of the daily require-
Soya bean oil/ Soya beans
Flax seed
Sweet peppers
Black salsify
Savoy cabbage
Avocado

A Vitamin E deficiency inevitably results in insufficient protection against the increased oxidative stress within the inflamed joint. This leads to an elevated destruction of cells and tissues (such as cartilage). The destroyed cells and the tissue in turn promote an increase of the inflammation. A vicious cycle begins. In a large number of scientific studies it was found, that the administration of vitamin E alleviated rest pains, chronic pains and movement pain, also mobility and increased grip strength improved, while morning stiffness was reduced and pain-free walking time was extended. The in those studies administered vitamin E dose was between 500 and 1200 IU day (α-tocopherol). The results of the studies were generally positive. A basic substitution of 300 to 600 IU of α-tocopherol is essential. A substitution in osteoarthritis of 800 to 1200 IU of α-tocopherol appears to be necessary and therapeutically effective.

Vitamin C and Vitamin A

Vitamin C protects vitamin E against oxidation. Oxidation causes detrimental change to cells, lipid-soluble vitamins, and polyunsaturated fatty acids via the influence of oxygen. The vitamin C in lemon juice, for example, inhibits the oxidation of apples. More precisely, it prevents the browning of apples. Every day at least 200 mg of vitamin C should be supplied to the body via food or tablets. The vitamin A levels in rheumatic patients are often decreased. Vitamin A substitution, however, cannot be recommended at this stage, based on the information available from current scientific studies.

Trace elements

Copper, selenium and zinc are components of a variety of enzymes and play an important role in the inflammatory response. Often in inflammatory rheumatic diseases levels of zinc, selenium and copper are lowered. During acute episodes 10 to 20 mg zinc can be substituted. The selenium intake (38 to 47 micrograms) in Germany is relatively low and is well below the recommended amount (50 to 100 micrograms). In active rheumatoid arthritis selenium substitution of 200 micrograms is highly recommended. To increase the bioavailability of zinc it should be supplied in organic form - for example, as 'Zinkorotat' or 'Zinkhistidin'.

Iron

Low blood iron levels can often be found in inflammatory diseases such as rheumatism. This can lead to anemia. Replacement therapy is, however, usually not indicated in inflammatory rheumatic diseases. If the physician diagnoses an iron deficiency, low levels of ferritin and high levels of transferrin, a compensation with iron supplements prescribed by the doctor is recommended. In these cases, copper should be administered as well.

> Note: Some of the most recent clinical studies have demonstrated that excessive food intake or even medical treatment can promote the inflammatory process. Before taking iron-containing supplements the doctor should be consulted.

Fish as a source for fatty acids

Fish fatty acids displace arachidonic acid from phospholipids. Similar to antioxidant agents they reduce the formation of eicosanoids by the inhibition of cyclooxygenase and lipoxygenase. In clinical studies, fish oil fatty acids turned out to be effective therapeutic agents in rheumatoid arthritis. The effect of eating food low in arachidonic acid is reinforced by the complementary administration of fish fatty acids.

> The recommended dose is 25 to 35 mg of omega-3 fatty acids per kg bodyweight (Example: 70 kg patient requires 1.75 to 2.45 grams of omega-3 fatty acids).

Several studies established a reduction of pain, reduced morning stiffness of the joints, greater grip strength and lower number of painful and swollen joints in patients eating a diet low in arachidonic acid additionally supplemented with fish oil capsules (drug quality!). The condition of the patients improved significantly and the medication was partially reduced. If the eating habits reverted back to the consumption of a conventional arachidonic-rich diet, the symptoms returned. The nutritional therapy in inflammatory rheumatic diseases has to be maintained permanently to be successful.

Dihomo-γ-linolenic acid and γ-linolenic acid

Some vegetable oils (soy, canola, wheat germ, walnuts and flaxseed) are the main sources for the essential omega-3 fatty acid α-linolenic acid, which once obtained via the diet, is converted to eicosapentaenoic acid. Eicosanoid synthesis is also inhibited. One of the positive effects of dihomo-γ-linolenic acid, is the displacement of arachidonic acid from the membrane lipids. The γ-linolenic acid precursor is absorbed by the organism in particular via evening primrose or borage oil. In the body, the γ-linolenic acid converts by chain extension to dihomo-γ-linolenic acid. A conventional diet supplies only very small amounts of the two fatty acids (only 0.01 to 0.02 g) to the body. Therapeutic effects like the inhibition of inflammation are, however, only to be expected if 2-3 grams are supplemented daily. The effect is similar to the one of eicosapentaenoic acid.

Lacto-vegetarian food including low-fat dairy products

Patients with inflammatory rheumatic diseases can actively and effectively support their therapy by following a lacto-vegetarian diet (milk, dairy products and vegan (vegetable) foods), which includes fish and is rich in calcium, antioxidant vitamins and omega-3 fatty acids. With such a diet the body is supplied with only 50 mg of arachidonic acid, while a conventional diet contains between 200 and 400 mg daily.

Oxidative stress

The use of vitamin supplements in the treatment of rheumatism has been repeatedly discussed and is considered to be useful. Especially vitamin E helps with arthritis and osteoarthritis. Through the process of aging and excessive usage, the joint wears out and is ultimately destroyed. According to the general theoretical view, this process is promoted by highly reactive molecules. They are produced as part of the cellular metabolism and oxygen consumption. It is believed that vitamin E is able to 'scavenge' and thereby eliminate these aggressive and harmful free radicals. Which is why vitamins E and C and the mineral selenium are also called antioxidants.

Vicious cycle oxidation:

- Vitamin E protects against the oxidation of arachidonic acid which yields inflammatory mediators like eicosanoids
- Vitamin C regenerates oxidised vitamin E
- A selenium-containing substance regenerates vitamin C
- A copper-containing substance protects it.

Therefore, rheumatic patients must ensure an optimal supply of vitamin E, vitamin C, selenium and copper. The vitamin C intake should be twice that of the vitamin E. Otherwise, an elevated vitamin E supply might promote pro-oxidant or pro-inflammatory processes. The formation of eicosanoid is an oxidative process, which can be inhibited by antioxidants and enzymes (e.g. proteins containing metal atoms). Oxidative stress has a negative effect on patients with inflammatory rheumatic diseases. A study demonstrated that oligoarthritic patients, who were smokers, twice as often developed a positive rheumatoid factor and joint destruction than nonsmoking subjects. Any inflammation is an oxygen-consuming process. The arachidonic acid contained in meat can only be converted into inflammatory mediators by means of oxygen radicals. Antioxidants (e.g. vitamin E) are able to protect unsaturated fatty acids from oxidation by oxygen radicals and thus inhibit the conversion of arachidonic acid to dangerous inflammatory mediators.

Recommended daily intake of micronutrients for healthy and rheumatic subjects		
Micronutrients	Healthy	Rheumatic patients
Vitamin A	1.8 mg	1.8 mg
Vitamin C	75 mg	800 mg
Vitamin E	12 mg	400 mg
Copper	1.5 mg	3 mg
Selenium	100 mg	200 mg
Zinc	15 mg	30 mg

Summary

Dietary therapy tailored to the needs of the patient reduces the level of inflammation in inflammatory rheumatic diseases. This is particularly dependent on the amount of arachidonic acid supply. Omega-3 fatty acids as well as omega-6 fatty acids like dihomo-γ-linolenic acid (DGLA) have an anti-inflammatory effect and are substituted on a daily basis. The diet tailored to the rheumatic patient is primarily a vegetable diet, supplemented with low-fat dairy products, fish and vegetable fats. Animal arachidonic-rich foods such as meat and sausages are largely excluded.

- Food low in arachidonic acid means to eat little meat and meat products.
- An omega-3 fatty acid-rich diet means to eat fish more often.
- More plant-based foods should be consumed, because they do not contain arachidonic acid, but instead plenty of vitamins and minerals.
- Whole-grain products and legumes cover the iron and selenium requirement.
- Only vegetable fats should be consumed, because they contain a lot of vitamin E and no arachidonic acid.
- Soybean oil should be used for cooking and yellow fat, rapeseed oil and rapeseed oil based margarines (rich in vitamin E).
- Low-fat dairy products should be consumed daily, because they contain plenty of calcium and prevent osteoporosis during treatment with cortisone.

Osteoarthritis

Osteoarthritis is a chronic degenerative rheumatologic joint syndrome of varied etiology. It is also conventionally known as 'arthrosis'. The aging of the articular cartilage affects the vascular permeability for nutrients and will result in a decreased amount of mucopolysaccharides, which altogether leads to the softening, cracking and erosion of the cartilage. The onset of osteoarthritis will continue to be promoted through all dysfunctions, which is why we call these factors also pre-osteoarthritic deformities. The most common clinical form of arthritis is osteoarthritis. It usually occurs in older people and mainly affects the joints of the lower extremities, such as the hip or knee, resulting in chronic diseases. In the later stages, this disease leads to the degeneration and destruction of the joint surfaces (articular cartilage and -bone). Usually, an artificial joint replacement is necessary at an advanced stage.

Degenerative joint disease is often caused by obesity, excessive loads, or poor posture. Other accompanying symptoms of degenerative joint disease in addition to diabetes mellitus and heart failure are hyperlipidemia, hyperuricemia and varicose veins (1). Especially older people may suffer from wear and tear of the joints, most commonly in the knee, hip and back joints. The cartilage layer caps the articulating surfaces of bones forming a synovial joint. The inner membrane of synovial joints secretes synovial fluid into the joint cavity, so that the cartilage is constantly supplied with nutrients. With increasing age, the cartilage layer loses its elasticity, turns fibrous and regresses. If cartilage is worn, the associated bone and the bones surrounding the joint capsule with its ligaments and muscles are also affected. Other causes of degenerative joint diseases are meniscus and cruciate ligament injuries sustained during competitive sports or poorly healed fractures. Osteoarthritis is classified either as primary, so-called idiopathic, or as secondary osteoarthritis. Primary osteoarthritis has been defined as an idiopathic condition developing in previously undamaged joints in the absence of an obvious causative mechanism. Genetic disposition, poor circulation caused by hormonal malfunctions or mechanical loads have been proposed to playing a crucial role in the development of osteoarthritis. Secondary osteoarthritis occurs frequently as a concomitant disease of other conditions, such as joint misalignments, non-physiological joint loads or due to obesity putting severe stress on the joints.

Metabolic disorders such as ochronosis, the accumulation of homogentisic acid in connective tissue, or diabetes mellitus also lead to secondary osteoarthritis. In addition to metabolic disorders, vascular complications and diabetic polyneuropathy play an important role in the development of osteoarthritis. Other metabolic causes are gout and calcium pyrophosphate crystal deposition (pseudogout), which are usually found in the meniscus and in intervertebral discs. Recurrent intra-articular bleeding is a major clinical manifestation in patients with hemophilia A and B and can also result in degenerative arthritis (so-called 'hemophilic joint'). Post-infectious arthritis often occurs after purulent arthritis, which destroys the cartilage. Examples of which are gonorrhea and tuberculosis. Other major causes are autoimmune diseases (rheumatoid arthritis, lupus erythematosus, and scleroderma) and aseptic osteonecrosis near the joint. No gender differences have been observed, however, hip and knee osteoarthritis are more prevalent in men, while women tend to suffer from osteoarthritis of smaller joints such as the fingers, where characteristic bony enlargements can form, the so-called Heberden's nodes (osteophytes).

Pathogenesis of osteoarthritis

The primary event that leads to osteoarthritis, is the breakdown of the articular cartilage via external or internal events. Examples are stress factors, metabolic disorders or infections. The chondrocytes, embedded in the intercellular substance of the cartilage, which are responsible for the maintenance of cartilage layer, secrete cytokines. These pro-inflammatory cell messengers of the immune system regulate the inflammatory response. The effect of the cytokines is increased by an increased sensitivity of the receptors in the arthritic joint. The destruction of cell membranes via the activity of phospholipases results in the release of phospholipids containing eicosanoids (leukotrienes and prostaglandins), inflammatory mediators. Released eicosanoids mediate vascular changes and changes in the vascular connective tissue, leading to clinical symptoms of osteoarthritis.

Morphology

The cartilage caps of the joint surfaces show an irregular and roughened contour when viewed under the microscope. Osteophytes (bony projections that form along joint margins) often develop on the edge of the cartilage caps, but mostly on the vertebral bone. Subchondral cysts are often found as large pseudozysts, filled with necrotic bone and bone fragments beneath the destroyed cartilage cap. Light-microscopically, asbestos-like fibrous striations, superficial defects and hyperplastic chondrocytes are visible in the hyaline cartilage. The subchondral bone, as a result of excessive regeneration, appears widened and clumped (cancellous bone), while the medullary cavity is fibrotic and contains necrotic cartilage fragments and micro fractured trabeculae. Histologically, the meniscus tissue shows a loosening of the ground substance and an irregular fiber structure. The surface is characterised by osteonecrosis and tears in the edge region with apparent proliferating hyperplastic cartilage cells.

Initial manifestations of the disease are stiffness and pain after prolonged exercise. Later, patients may suffer from severe pain, which eventually can lead to movement disabilities. This pain can occur when resting, at night or can be caused by the weather. Diagnosis of osteoarthritis is not always easy at first, because although osteoarthritis is clinically detectable, it is often silent, painless and therefore not active. Often there are discrepancies between pathological, radiological detectable and clinically relevant types of osteoarthritis. In case of suspected osteoarthritis, X-ray computed tomography, magnetic resonance imaging and arthroscopy are indicated. Only once these diagnostics are completed, the patient will receive medical treatment, as only 20 to 30% of osteoarthritic diseases being painful (1).

Osteoarthritis can affect any joint. The most common types of arthritis are:
- Osteoarthritis of the knee (Gonarthrosis):
  Arthritic changes predominantly affect the medial or lateral articular surface depending on the static load.
- Osteoarthritis of the hip (Coxarthrosis)
  An increase in deformity may lead to subluxation (partial joint dislocation)
- Osteoarthritis of the upper and lower ankle joint
- Osteoarthritis of the thumb joint (Rhizarthrosis)
- Osteoarthritis of the shoulder joint (Omarthrose)
  Joint space narrowing
- Osteoarthritis of the spine (Spondylarthrosis)
  Joint space narrowing
- Osteoarthritis of the distal interphalangeal joints (Herberden-Arthrose)
- Osteoarthritis of the proximal interphalangeal joint (Bouchard-Arthrosis)
- Osteoarthritis of the carpometacarpal joint (Rhizarthrose)
- Osteoarthritis of the metacarpophalangeal joint (Hallux rigidus)
- Misalignment of the big toe (Hallux valgus)
- Osteoarthritis of the sacroiliac joint
- Osteoarthritis of the temporomandibular joint
- Osteoarthritis affecting multiple joints (Polyarthrosis)

Inflammatory joint disease include arthritis, rheumatism, rheumatoid arthritis and their special subtypes, psoriatic arthritis, ankylosing spondylitis and joint disease after infections, such as urinary tract or bowel inflammation (2, 3, 4).

Treatment

There is no cure for osteoarthritis, although a variety of cartilage supplements are available. These range from gelatin to herbal substances with different ingredients. So far, scientific evidence proving their effectiveness is still lacking. However, various preventative measures may result in a significant relief of osteoarthritic symptoms:
- **Weight reduction** when being overweight.
- Effective **Physiotherapy** and **therapeutic exercise.**
- **Intra-articular joint injections** with flushing of the joint and injection of cortisone in inflammatory phases of osteoarthritis or application of local anesthetics as pain therapy.
- Injection of hyaluronic acid into the knee joint, which acts as a 'joint lubricant' and some patients benefit from longer pain relief.
- **Orthopaedic technology** (stick, buffered soles, raising the inner or exterior shoe ridge).
- **Painkillers**: e.g. cyclooxygenase-inhibitor.

The treatment of osteoarthritis is based on three objectives: the maintenance of joint function, pain reduction, and to stop the destruction of cartilage. Often, weight loss, strengthening of the joint-encompassing muscles, and uniform, low-stress movement of the joints are effective first-aid measures before the implementation of medical treatment. Weight loss reduces the static pressure on the cartilage when standing upright. The strengthening of the joint-encompassing muscles, reduces pain and improves joint function. Cyclic, uniform and low-stress movements such as cycling positively affect even advanced osteoarthritis. The Drug

Commission of the German Medical Association (AkdÄ) recommends the administration of analgesics for the treatment of osteoarthritis. These include, depending on the degree of pain, nonsteroidal anti-inflammatory drugs (NSAIDs) such as paracetamol, diclofenac or ibuprofen, easy on the stomach cyclooxygenase-2 inhibitors (COX) inhibitors and potent analgesics of the morphine-type. They inhibit the formation of inflammatory cytokines and metalloproteinase and stimulate the proteoglycan synthesis in chondrocytes.

Some pain killers are suspected to destroy the cartilage long-term and prevent tissue neogenesis. An alternative are so-called chondroprotectives. These are combination products with chondroitin sulphate, glucosamine and hyaluronic acid respectively their derivatives. In the following, we will focus on studies of glucosamine and chondroitin.

General nutrition guidelines:

General recommendations are to maintain a normal body weight and a balanced, vitamin and mineral-rich diet. The latter should be rich in fruits and vegetables, whole grains, bone-fortifying calcium and vitamin D, vitamin E and C as radical scavengers and selenium. In addition, a healthy lifestyle should be maintained, which means to avoid smoking. For smoking narrows vessels and reduces the oxygen supply to the body, and thus to the joints.

What are glucosamine and chondroitin?

Connective tissue is the most widespread and abundant type of tissue in the human body. Each tissue is typically composed of cells and extracellular fibres, the latter being the chief component of the connective tissue mass. The cells of our body synthesise the extracellular and cell-associated components from low molecular weight substrates. The fibroblast is responsible for the synthesis of the fibrous proteins collagen, elastin, and for the proteoglycans making up the extracellular matrix of the connective tissue. The body's building blocks glucosamine (glucosamine sulphate) and chondroitin (chondroitin sulphate) are components of all major body tissues, even the blood vessels and heart valves. The main function of glucosamine and chondroitin sulphate is development and repair of tissue of the musculoskeletal system. Connective tissue, ligaments, tendons, cartilage, bone and joint health with enough synovial fluid are dependent on an adequate supply of these substances. If the body is not enough glucosamine available, the synovial fluid becomes thinner and more watery. The joints are then more prone to wear and injury. In other wear parts, such as the intervertebral discs, the situation is similar. The more glucosamine is available to the body the more he can produce cartilage. Normally the body produces enough glucosamine to keep the joints working and to repair minor damage. With age, the body's natural glucosamine production takes off, however, the joint dries out and the cartilage is poorly nourished, built up unevenly and minor injuries not heal.

Chemical structure of glucosamine and chondroitin

Glycosaminoglycans (GAGs) are highly polymeric compounds, called mucopolysaccharides, which include glucosamine, chondroitin and hyaluronic acid. The glucosamine molecule is smaller than chondroitin, which facilitates gastrointestinal absorption. The chemical bond between glucosamine and chondroitin sulphate with the salt of sulfuric acid results in the formation of the bioactive forms of glucosamine and chondroitin which are glucosamine sulphate (Fig. 1) or chondroitin sulphate (Fig. 2). Only those substances can be absorbed and metabolised by the organism. The amino sugar D-glucosamine (2-amino-2-deoxyglucose) is the natural component of many oligo- and polysaccharides. Glucosamine is involved in the biosynthesis of amino sugars and is converted to glucosamine-6-phosphate under ATP release. As a final product result UDP-N-acetylglucosamine, N-substituted amino sugar (N = nitrogen), and UDP-N-acetylgalactosamine. UDP-N-acetylglucosamine are involved in the further course of the synthesis of glycosaminoglycans and glycoproteins.

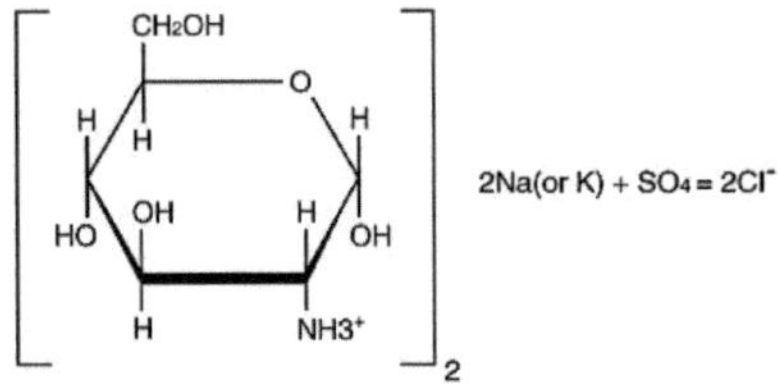

Figure 1: Glucosamine sulphate

Glucosamine is most abundant and contributes to the creation of all buffer layers in the body. That means that glucosamine is involved in the development or growth of cartilage in the joints and the synovial fluid, the essential protectors of the joint system. 50 % of the synovial fluid ('joint lubricant') is composed of hyaluronic acid, a mucopolysaccharide, which is a precursor of glucosamine. This intracellular cement is an important component of the connective tissue. Insulin deficiency and regular administration of corticosteroids disturbs or prevents the body's own hyaluronic acid production. Hyaluronic acid gives synovial fluid its toughness and lubricates the surfaces between synovium and cartilage.

Chondroitin sulphate is a sulphated glycosaminoglycan. It is composed of a chain of alternating sugars covalently attached to core proteins. These anionic linear polymers contain either of two modified sugars - N-acetylgalactosamine or N-acetyl glucosamine - and an uronic acid such as glucuronate or iduronate (5). After oral ingestion, the molecule is enzymatically cleaved into mono- and disaccharides, which can cross the intestinal wall. The cleavage products coalesce with the glycosaminoglycans. They represent the basis for the synthesis of glycosaminoglycans in the cartilage, which are responsible for shock absorbing properties and friction-free movement of the cartilage. Due to its structure, chondroitin sulphate has the highest water binding capacity of all sulphated glycosaminoglycans.

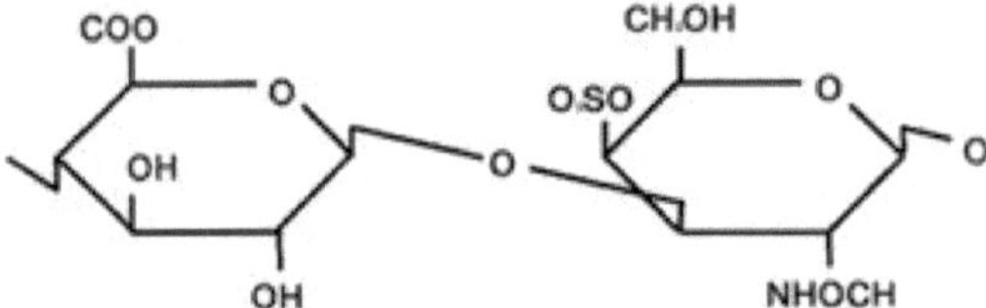

Figure 2: Chondroitin sulphate

The role of glucosamine and chondroitin in the human body

Proteoglycans are synthesised in all cells at different concentrations. Below some of their functions are listed:
- Mediation of cell-cell contacts and cell-matrix interactions through cell membrane-integrated proteoglycans.
- Control of cell proliferation by specific binding of growth factors.

- Viscoelastic behavior of high molecular weight hydrated proteoglycans under pressure load.

Glucosamines are thought to be involved in cartilage repair and cartilage and bone formation. Chondroitin is a cartilage component that promotes water retention and elasticity and protects other endogenous enzymes against the weakening of cartilage components. It is usually present in cartilage rich in water (5). Effective cartilage protective agents should be able to perform the following tasks:

- Enhance chondrocyte protein synthesis (glycosaminoglycans, proteoglycans, collagen, proteins, RNA and DNA).
- Promoting the synthesis of hyaluronic acid.
- Inhibition of cartilage-degrading enzymes.
- Activation of platelets, fibrin, lipids and cholesterol in the interstices of the synovial membrane.
- Joint pain relief.
- Control inflammatory synovitis.

The interaction and synergetic effect of joint nutrients heals the cartilage matrix and positively affects cartilage self-repair on a cellular level.

Study results

Research studies suggest that glucosamine promotes the production of cartilage-building proteins. Other studies find that chondroitin inhibits the production of cartilage-destroying enzymes and fights inflammation too. Supplements extracted from crustacean shells and bovine cartilage and used in human clinical studies show better efficacy than conventional arthritis medications. Feeling less stiff, plus less side effects and reduction of pain are the main benefits (6). Clinical studies often use the Lequesne Algofunctional Index and also the WOMAC Index to assess the severity of the osteoarthritis. The Lequesne Index is a standard point-value, which can be used by the patient to describe the pain, walking ability and the functional limitations of osteoarthritis. The WOMAC Index is a self-questionnaire to be filled out by the patient, to score their pain, joint stiffness and physical functionality.

In the randomised, placebo-controlled and double-blind study by Noack *et al.* (7), 252 patients were treated daily for four weeks with 500 mg of glucosamine sulphate or a placebo. Initially, the Lequesne Index in both groups was 10.6 and decreased to 7.45 points in the 'glucosamine group'. The placebo group had a mean Lequesne Index of 8.4 points. The medication was well tolerated by both groups. Reginster *et al.* (8) and Pavelka *et al.* (9) independently determined over a period of three years in randomised, double-blind, placebo-controlled clinical trials the effect of glucosamine sulphate in osteoarthritis. Reginster *et al.* studied 212 patients with primary osteoarthritis of the knee. The patients had a BMI below 30 and were over 50 years old. They received daily 1500 mg glucosamine sulphate or a placebo for three years. In contrast to the placebo group, the group treated with glucosamine showed no significant changes in joint space width (glucosamine group: 0.07 mm; 95 % confidence interval, -0.17 to 0.32 mm. Placebo group: -0.31 mm; 95 % confidence interval, -0.57 to -0.04 mm). The WOMAC Index changed in the group treated with glucosamine after three years to -24.3 % (95 % confidence interval, -37.0 to -11.6 mm). In the placebo group, a change of 9.8 % (95 % confidence interval, -14.6 to 34.3 mm) was recorded. As a result, it was confirmed that an application with glucosamine seems to relieve mild to moderate subtypes of osteoarthritis. Comparing the placebo with the glucosamine group, it is clear that the glucosamine group

showed no progression of clinical symptoms and even relief of clinical symptoms. However, no difference in joint stiffness was ascertained (8). Both groups showed no side effects.
Pavelka *et al.* studied 202 patients for three years with mild to moderate osteoarthritis of the knee, during which the influence of 1500 mg glucosamine sulphate or a placebo daily was tested and found to delay the development of osteoarthritis of the knee. X-ray revealed little, discernible changes in the knee joint gap width measurements. The average joint space width was less than 4 mm. After the study was completed, a progressive narrowing of the joint space by 0.19 mm (95 % confidence interval, -0.29 to -0.09 mm) was recorded in the placebo group. As a result of the administration of glucosamine sulphate the joint gap did not significantly change (0.04 mm, 95 % confidence interval, -0.06 to 0.14 mm). The difference between those two groups was therefore significant. Prior to the study, the as severe narrowing of the joint space (0.5 mm) defined osteoarthritis of the knee, was less frequent (5 % as opposed to 14 %) in the group treated with glucosamine. At the same time, the symptoms improved by 20 to 30 % compared to the placebo group. After completion of the study the two groups differed significantly in the Lequesne and WOMAC Indices regarding physical impairment and stiffness. Finally, Pavelka and his team found, that long-term treatment with glucosamine sulphate slowed down the progression of knee osteoarthritis, implying that this substance potentially has a disease-modifying effect (9). The treatment in both groups showed no side effects.

Eaton and Dr. Lloyd found that chondroitin sulphate together with glucosamine improves the formation of new cartilage. Chondroitin inhibits proteoglycans-degrading enzymes and, in addition stimulates the synthesis of proteoglycans and collagen (10). Bucsi and Poor (11) investigated the efficacy and compatibility of oral chondroitin sulphate in osteoarthritis of the knee. In the randomised, double-blind, placebo-controlled study for six months twice daily 400 mg chondroitin sulphate or a placebo was administered orally to the study participants. At the beginning of the study, after one, three and six months the 80 patients were clinically examined. In summary, it can be said that the 'chondroitin sulphate group' could walk longer without pain than the placebo group. Patients with placebo consumed slightly more paracetamol. These results suggest that chondroitin sulphate is a slow-acting drug suitable to treat osteoarthritis of the knee.

Reichelt *et al.* (12) studied the efficacy of twice-weekly 400 mg intramuscularly injected glucosamine sulphate for six weeks in their randomised, placebo-controlled, double-blind study of 155 patients. The subjects in both groups had suffered for at least six months from osteoarthritis of the knee and initially had a Lequesne Index of an average of ten points or a little bit more. After taking into account all cases that did not complete the study, 51 % of the 'glucosamine-group' (equivalent to 73 subjects) reported significant improvements, in contrast to 30 % of the placebo group (69 subjects). Muller-Fassbender *et al.* (13) found in their randomised, double-blind study of 200 subjects with osteoarthritis of the knee, that a three times daily oral administration of 500 mg of glucosamine sulphate versus 400 mg ibuprofen during four weeks, reduced the Lequesne Index from more than 12 points to 10. At the same time, no difference between the glucosamine and the ibuprofen group was observed. The Lequesne Index was on average 16 points and decreased in both groups by at least six points. In contrast to the 'glucosamine group' 35 % of the 'ibuprofen group' suffered from gastrointestinal side effects. In conclusion, this suggests that glucosamine sulphate is comparable to ibuprofen in its anti-inflammatory efficacy, which is why glucosamine sulphate can be recommended as a safe and slow-acting drug for osteoarthritis of the knee.
Studies have shown that the glucosamine does not only relieve symptoms, but that it can also prevent the progressive loss of cartilage caused by osteoarthritis. In 1999 Reginster and his team examined the efficacy of glucosamine sulphate 1500 mg daily in a three-year controlled

study of 212 patients diagnosed with osteoarthritis of the knee. In the group receiving this drug, the narrowing of the joint space increased only a fraction (0.06 mm), in the placebo group however, the gap was on average 0.31 mm narrower. In addition, in this group the symptoms of osteoarthritis (stiffness, loss of function, pain) increased, while these symptoms decreased in the other patients (14).

How to prevent osteoarthritis

Glucosamine and chondroitin sulphate are structural components of cartilage in animal tissues and thus, it is difficult to supply the human body with these substances. It is therefore advisable to take preventive measures to protect the joints. Weight reduction, moderate sport and the correction of a poor posture are just a few ways to counteract osteoarthritis. Glucosamine and chondroitin are suitable drugs to treat arthritis. If the supplementation of the latter substances does not result in a noteworthy improvement of the symptoms, the patient may still resort to conventional medications. Strategies for the prevention of osteoarthritis:
- Avoid walking on uneven paths (impact load)
- Sensible alteration of loading and unloading of joints
- Use of walking aids when indicated
- Use of shoes with buffered heels (soft soles)
- Avoid cold and damp
- Keep the joints warm (microclimate)
- Casual gymnastics
- Swimming in warm water

The role of nutritional therapy in degenerative joint diseases

The effective treatment of osteoarthritis targets four basic mechanisms. Firstly, the reduction or inhibition of inflammation of the affected joints on the cellular level and also the modulation of the synthesis of proteoglycans and hyaluronic acid, major components of the cartilage matrix. Direct protective features are the inhibition of proteolytic enzymes and the reduction of damage to the matrix molecules through free radicals. At least as important as the previously mentioned goals of treatment, is the protective effect on the cellular cartilage components. Evidentially, chondroitin delays and protects against cartilage degradation. Glucosamine is an important building block of the cartilage matrix and amongst other things inhibits catabolic processes in the articular cartilage by inhibiting elastase release from activated granulocytes as well as the expression of chondrocytes-derived aggrecanases (15, 16, 17). Regular treatment with glucosamine and chondroitin has an increased latency of about six to eight weeks before the therapeutic effect emerges. They work slowly. However, in terms of clinical efficacy, they are no different to non-steroidal anti-inflammatory drugs (NSAIDs). A meta-analysis confirmed the effect described (18). However, the patient can often only confirm definite effects after taking the medication for a few months. Another important benefit of the treatment with glucosamine and chondroitin is the lasting effect over several months after discontinuation of the medication. This allows two annual cycles of therapy, during which chondroitin and glucosamine are prescribed for three months. At the same time, compared to conventional drugs used in the treatment of osteoarthritis, glucosamine and chondroitin are known to have substantially reduced side effects. Only some patients sporadically report slight stomach upset, flatulence and softer stools. COX-inhibitors have been associated with cardiovascular complications and non-steroidal anti-inflammatory drugs (NSAIDs) are known to have very strong gastrological side effects. Animal experiments indicate that there is a probability that glucosamine could increase the insulin resistance in type 2 diabetics. Studies based on human subjects could not detect these risks so far. Patients with diabetes mellitus should critically monitor their blood sugar levels if using these drugs for treatment is contemplated. So far, there are no signs of allergic reactions when taking glucosamine. As it is, however, derived from shellfish, patients who are allergic to the latter should keep a close eye on the development of any side effects or completely refrain from taking these drugs. There is an increased risk in hemophiliacs or patients who are taking blood thinners, that chondroitin sulphate may induce bleeding. From a medical perspective, the nutritional supplementation of glucosamine and chondroitin is useful, if patients are simultaneously supplemented with vitamins C and E because of their anti-inflammatory and strengthening effect on the connective tissue; selenium to catch free radicals; copper to facilitate oxygen transport, folate to help form red blood cells; and zinc to strengthen the immune system.

Author:

Sven-David Müller, M.Sc.
Master of Science in Applied Nutritional Medicine, a nationally recognised dietician, advisor of the German Diabetes Association (DDG) and medical journalist, Haddamshäuser Weg 4a, 35096 Weimar an the Lahn, www.svendavidmueller.de, diaetmueller@web.de